Mohammad Shahjahan

Durable Solutions for Climate Displaced Persons

Mohammad Shahjahan

Durable Solutions for Climate Displaced Persons

Toward the durable solutions of climate displaced persons: An initiative of Bangladesh

LAP LAMBERT Academic Publishing

Impressum / Imprint

Bibliografische Information der Deutschen Nationalbibliothek: Die Deutsche Nationalbibliothek verzeichnet diese Publikation in der Deutschen Nationalbibliografie; detaillierte bibliografische Daten sind im Internet über http://dnb.d-nb.de abrufbar.

Alle in diesem Buch genannten Marken und Produktnamen unterliegen warenzeichen-, marken- oder patentrechtlichem Schutz bzw. sind Warenzeichen oder eingetragene Warenzeichen der jeweiligen Inhaber. Die Wiedergabe von Marken, Produktnamen, Gebrauchsnamen, Handelsnamen, Warenbezeichnungen u.s.w. in diesem Werk berechtigt auch ohne besondere Kennzeichnung nicht zu der Annahme, dass solche Namen im Sinne der Warenzeichen- und Markenschutzgesetzgebung als frei zu betrachten wären und daher von jedermann benutzt werden dürften.

Bibliographic information published by the Deutsche Nationalbibliothek: The Deutsche Nationalbibliothek lists this publication in the Deutsche Nationalbibliografie; detailed bibliographic data are available in the Internet at http://dnb.d-nb.de.

Any brand names and product names mentioned in this book are subject to trademark, brand or patent protection and are trademarks or registered trademarks of their respective holders. The use of brand names, product names, common names, trade names, product descriptions etc. even without a particular marking in this work is in no way to be construed to mean that such names may be regarded as unrestricted in respect of trademark and brand protection legislation and could thus be used by anyone.

Coverbild / Cover image: www.ingimage.com

Verlag / Publisher:
LAP LAMBERT Academic Publishing
ist ein Imprint der / is a trademark of
OmniScriptum GmbH & Co. KG
Heinrich-Böcking-Str. 6-8, 66121 Saarbrücken, Deutschland / Germany
Email: info@lap-publishing.com

Herstellung: siehe letzte Seite /
Printed at: see last page
ISBN: 978-3-659-64031-5

Toward the durable solutions of climate displaced persons: An initiative of Bangladesh

Mohammad Shahjahan

Dedication

I like to dedicate my article to the people of Bangladesh who have become displaced as an impact of climate change and been struggling to survive. A special feeling of gratitude to my colleague, Md. Arifur Rahman and Prabal Barua of YPSA, who have supported me throughout the writing process.

I would like to thank Scott Leckie and Ezekiel Simperingham of Displacement Solutions for inspiring me with their expertise and different writings. Finally, I express my heartfelt gratitude to my pretty wife (Moli) and two loving sons (Adib and Ariq) for allowing me to do this writing in late night.

Contents

Abstract..3

Introduction...5

Bangladesh Housing, Land and Property Rights Initiative 7

Housing, land and property rights under international law 7

The Domestic and International Legal Standard for Government of Bangladesh to address the rights of Climate displaced persons: 9

Existing Policies and Laws in Bangladesh Relevant to Climate Change and Climate Displacement 11

Report on Climate Displacement in Bangladesh: *The Need for Urgent Housing, Land and Property Rights Solutions*.. 29

Study on Land Availability for Climate Displaced Communities in Bangladesh... 30

Study on Land Acquisition for Climate Displaced Communities in Bangladesh ... 31

Study on The Viability of the Chittagong Hill Tracts as a Destination for Climate Displaced Communities in Bangladesh 32

Climate Displacement in Bangladesh: Stakeholders, Laws and Polices - Mapping The Existing Institutional Framework 33

Formulation of Peninsula Principles on Climate Displacement 35

Advocacy and Capacity Building on the Rights of Climate Displaced Persons in Bangladesh.. 39

Coastal Kids Project... 41

Development of Guidance Note of Access to *Khas* Land for Climate Displaced Persons in Bangladesh 41

Five Key Actions for ensuring the rights of climate displaced persons.................. 43

Concluding Remarks... 48

Recommendations:... 48

1. Undertake a comprehensive review of existing national laws and policies:.. 48

2. Design and implement rights-based national laws and policies: 49

3. Undertake a review and re-design of the existing institutional framework: .. 49

4. Design and implement domestic land solutions:... 49

5. Increase coordination and knowledge sharing among international and regional donors: 50

6. Create a UN and NGO working group on climate displacement: 50

7. Improve coordination and communication between the Government and civil society:.. 51

References.. 52

Abstract

In recent years climate change induced displacement has got vast importance as one of the major penalty of climate change impact. IPCC fifth assessment report (WGI) reveals that climate change is now indisputable, and since the 1950s, many of the observed changes are unprecedented over decades to millennia and entire globe is facing this as one of the main challenges. Climate displacement is already underway in many countries such as Bangladesh and expected to affect many tens of millions of people in coming decades. Climate change is likely to increase the intensity and severity of many of the natural hazards already suffered by Bangladesh as well as negatively impact on the wider socio-economic situation, creating widespread losses of housing, land and livelihoods and driving new surges of migration and displacement.[1] Many of these hazards are expected to disproportionately affect the poorest and most vulnerable in Bangladesh, a country where more than 50 million people still live in poverty. Among the 64 districts of Bangladesh, 26 coastal and mainland districts are already producing climate displaced people. But there is no comprehensive national policy in Bangladesh that specifically targets climate displacement. Resettlement for the displaced peoples should be a last resort in climate change adaptation, but the reality is that it is already occurring in some countries and this trend is likely to intensify. The climate-induced migrants are often discriminated and face different problems during or after the displacement. In many cases, the policies and institutional frameworks are not sufficient to protect the displaced persons. The rehabilitation of displaced persons by government and non-government sectors are, to date, insignificant in terms of the total number of displaced persons. Importantly, livelihood problems remain after the rehabilitation of displaced persons. There should be a rights-based solution to this problem. There should be initiatives to generate sustainable employment opportunity for the rehabilitated climate displaced persons. Therefore, there is a need to review the relevant policies and institutional frameworks identifying the protection gaps and adopting new policies to protect the environmental migrants. To find out the durable solutions of climate displacement in Bangladesh, Young Power in Social Action (YPSA), a Bangladeshi national NGO, has been implementing Bangladesh Housing, Land and Property (HLP) rights initiative with the support of Displacement Solutions (DS), a Switzerland based international NGO, since 2011. The main objective of this initiative is to identify and implement rights-based solutions and actions to resolve climate displacement across the country as well as to ensure and safeguard housing, land and property rights. In this paper, the author reviews the climate displacement and people's vulnerability, existing laws and policies, activities performed under the Bangladesh HLP initiative and finally suggests five key actions for rights based solution of climate displaced people of Bangladesh. The author also reviews the existing laws and policies in Bangladesh relevant to climate displacement and demonstrates four studies conducted jointly by YPSA and DS, of which three land studies that examined in detail the true picture of land availability in Bangladesh for climate displaced persons as well as the potential durability of future resettlement and one comprehensive mapping study on "Climate Displacement in Bangladesh: Stakeholders, Laws and Policies - Mapping the Existing Institutional Framework". The author highlights The Peninsula Principles, new international rights standard for climate displaced persons, which provide a comprehensive normative framework, based on principles of international law, human rights obligations and good practice, within which the rights of climate displaced persons can be addressed within a State and not cross-border climate displacement.

Key Words: Climate Displacement, Displaced Persons, Housing, Land and Property Rights, Durable Solutions, Bangladesh, Displacement Solutions, YPSA

Introduction

Warming of the climate system is unequivocal, and since the 1950s, many of the observed changes are unprecedented over decades to millennia[2] and entire world is facing these changes as one of the main challenges. Climate change impact in Bangladesh have been more worsening as the most vulnerable country of world due to its geographical location and flat topography. Bangladesh has ranked fifth among 10 countries most vulnerable to climate change-induced natural disasters in the last two decades from 1993 to 2012.[3] Assessing the Costs of Climate Change and Adaptation in South Asia report commissioned by ADB revealed that average temperature of earth may increase up to 2^0C by 2050, which can cause of economic loss of annual GDP of Bangladesh equivalent to 2% by 2050 and 9.4% by 2100.[4] Displacement due to the effect of climate change happening in many countries like Bangladesh and affecting millions of people. A number of scientific reports confirmed that a large number of people have already been displaced in Bangladesh because of the direct effects of climate change, and the number is likely to increase considerably in the future.[5] Climate displacement is not just a phenomenon to be addressed at some point in the future, it is a crisis that is unfolding across Bangladesh now. Sea-level rise and tropical cyclones in coastal areas, as well as flooding and riverbank erosion in mainland areas, are already resulting in the loss of homes, land and property and leading to mass displacement. Further, all of the natural hazards that are causing displacement are expected to increase in both frequency and intensity as a result of climate change – almost inevitably leading to the displacement of many millions more across Bangladesh.

The effects of climate change are expected to exacerbate many of these existing hazards, as well as create new drivers of displacement. Among the 64 districts of Bangladesh, 26 coastal and mainland districts are already producing climate displaced people. CDMP II report 2014 under the Ministry of Disaster Management and Relief of Bangladesh shown that about 46% people temporarily displaced and about 12% people permanently displaced due to different hazards like floods, river bank erosion, salinity and water logging in four climate hotspots of Bangladesh.[6] On the other hand due to impact of climate change, sizes of Kutubdia island reduced almost 50% in last 20 years. Since 1991, six villages on the island completely have gone to the sea and about 40,000 people have fled and most of them got temporary shelter to the coast near Cox's Bazar district. The same scenario can be found in the others Island of Bangladesh like Sandip island, Bhola island and Dhalgata union of Moheskhali Island, which has been going to small in size and people of those island losing their house, land and property and finally becoming displaced.

Study also revealed that almost 40 million people live in the coastal areas of Bangladesh, Loss of coastal land to the sea, currently predicted to 3% by the 2030s and 6% in the 2050s, is likely to generate steady flow of displaced people.[7] The coastal areas are particularly vulnerable to tropical cyclones and associated storm urges. In 1991 Cyclone, around 4% population of the country and 15% population of the coastal area were displaced from their homes.[8] Due to its unique geographic position Bangladesh suffers from regular natural hazards, including floods, tropical cyclones, storm surges and droughts that lead to loss of life, land, homes, livelihoods and to the forced displacement of individuals and communities across the country. However, through a combination of lack of political will as well as a lack of financial and technical resources, there are currently no comprehensive mechanisms to provide support to people who have lost their homes, land and property as a result of climate change to rebuild their lives. In many cases, the policies and institutional frameworks are not sufficient to protect the displaced persons. The climate-induced migrants are often discriminated and face different problems during or after the displacement. All persons displaced by climate change in Bangladesh are rights-holders and are entitled to respect and protection of the full range of their human rights under national and international law.

The rehabilitation of displaced persons by government and non-government sectors are, to date, insignificant in terms of the total number of displaced persons. Importantly, livelihood problems remain after the rehabilitation of displaced persons. There should be a rights-based solution to this problem. There should be initiatives to generate sustainable employment opportunity for the rehabilitation of climate displaced persons. Therefore, there is a need to review the relevant policies and institutional frameworks identifying the protection gaps and adopting new policies to protect the environmental migrants. To face the challenge of mass displacement (both internal and external) as a result of climate Change, Young Power in Social Action (YPSA), a social development organization, has been implementing Bangladesh Housing, Land and Property (HLP) rights initiative with the support of Displacement Solutions (DS) for rights based solutions of climate displaced persons since 2011. The main objective of this initiative is to identify and implement rights-based solutions and actions to resolve climate displacement across the country as well as to ensure and safeguard housing, land and property rights. With this initiative, DS and YPSA have prioritized the question of how to resolve climate displacement in a rights-based manner that recognizes and protects the broad spectrum of housing, land and property (HLP) rights of climate displaced persons across Bangladesh. And initiative has also

- Published a series of ground-breaking reports on climate displacement in Bangladesh including Climate Displacement in Bangladesh: Stakeholders, Laws and Policies- Mapping the Existing Institutional Framework, which have been influential in national and international policy discussions on climate displacement;

- Prioritized the question of acquiring land for the relocation of climate displaced persons from unsafe areas to areas of safety from climate induced hazards. This has included the completion of three comprehensive research reports, identifying land available for relocation as well as the legal and regulatory process by which climate displaced persons can acquire such land;

- Raised awareness of approximate 1000 civil society representatives, from the grassroots to the national level, on the human rights of climate displaced communities in Bangladesh; and

- Undertaken advocacy, developing advocacy tool with five key actions, on the rights of climate displaced persons with key Government officials, including standing Ministers of the Government of Bangladesh and leaders of three primary political parties in Bangladesh.

Bangladesh Housing, Land and Property Rights Initiative

Displacement Solutions and YPSA have initiated the Bangladesh Housing, Land and Property (HLP) Rights Initiative in 2011 to improve the prospects of people and communities displaced by climate change. This Initiative has undertaken different studies that examined in detail the true picture of climate displacement including intensive advocacy and lobbying, both on raising the profile of climate displacement and increasing political awareness of this issue.

Housing, land and property rights under international law[9]:

Significant development of housing, land and property rights has occurred at the international level over the past 50 years. This development began with the 1948 Universal Declaration of Human Rights (UDHR), which recognized both the right to housing and the right to property. Since that time, HLP rights have been reaffirmed and developed in a series of international human rights treaties, declarations and other documents. There has also been a number of interpretative standards and documents developed at the international level, including the UN Committee on Economic, Social and Cultural Rights General Comment No. 4 on the Right to Adequate Housing, General Comment No. 7 on Forced Evictions, the Guiding Principles on the Rights of Internally Displaced Persons and the Pinheiro Principles on Housing and Property Restitution for Refugees and Displaced Persons.

Combined, these legal sources create a considerable body of international human rights laws and standards relevant to climate displaced people in Bangladesh. In broad way HLP rights as follows:

- The right to adequate housing and rights in housing;
- The right to security of tenure;
- The right not to be arbitrarily evicted;
- The right to land and rights in land;
- The right to property and the peaceful enjoyment of possessions;
- The right to privacy and respect for the home;
- The right to HLP restitution/compensation following forced displacement;
- The right to freedom of movement and to choose one's residence;
- The right to political participation;
- The right to information;
- The right to be free from discrimination;
- The right to equality of treatment and access;
- The right to water; and
- The right to energy.

All climate displaced people in Bangladesh are rights-holders and have the right to claim respect for these standards from the Government of Bangladesh. Of all of these rights, the right to adequate housing has been developed the farthest at the international level. In 1991, the UN Committee on Economic, Social and Cultural Rights adopted 'General Comment No. 4 on the Right to Adequate Housing' which indicates that the following seven components form the core contents of the human right to adequate housing:

8

- Legal security of tenure;
- Availability of services, materials, facilities and infrastructure;
- Location;
- Habitability;
- Affordability;
- Accessibility; and
- Cultural adequacy.

To promote HLP rights, Bangladesh should adopt targeted measures such as national HLP strategies that explicitly define the objectives for the development of the HLP sector, identify the resources available to meet these goals, the most cost-effective way of using them and how the responsibilities and time-frame for their implementation will be applied. Such strategies should reflect extensive genuine consultation with, and participation by, all those affected, including groups traditionally excluded from the enjoyment of HLP rights.

The relationship between housing, land and property rights and climate displacement in Bangladesh is clear –the current primary causes of displacement in Bangladesh are the destruction of homes and the loss of land as a result of climate hazards. The legally binding obligations on the Government to respect, protect, promote and fulfil the housing, land and property rights of climate displaced persons must be used as the basis for legislative, policy and institutional solutions to climate displacement. Adopting housing, land and property rights orientated policies and programmes for climate displacement in Bangladesh will form the basis for truly durable solutions for millions of climate displaced people now and in the future.

The Domestic and International Legal Standard for Government of Bangladesh to address the rights of Climate displaced persons: The Government of Bangladesh has clear responsibilities under domestic and international law to provide rights-based – and particularly housing, land and property rights - solutions to climate displaced persons. The Constitution of the People's Republic of Bangladesh states:

Article 27 *"All citizens are equal before law and are entitled to equal protection of law"*.

Article 15 *"It shall be a fundamental responsibility of the State to attain, through planned economic growth, a constant increase of productive forces and a steady improvement in the material and cultural standard of living of the people, with a view to securing to its citizens (a) the provision of the basic necessities of life, including food, clothing, shelter, education and medical care; (b) the right to work, that is the right to guaranteed employment at a reasonable wage having regard to the quantity and quality of work"*

Article 18A *"The State shall endeavor to protect and improve the environment and to preserve and safeguard the natural resources, bio-diversity, wetlands, forests and wild life for the present and future citizens"*

Article 19(2) *"The State shall adopt effective measures to remove social and economic inequality...and to ensure the equitable distribution of wealth among citizens, and of opportunities in order to attain a uniform level of economic development throughout the Republic".*

Article 25 *"The State shall base its international relations on the principles of respect for national sovereignty and equality...and respect for international law and the principles enunciated in the United Nations Charter".*

Besides, Bangladesh has signed and is bound to respect many key international human rights treaties that provide important human rights protections to climate displaced persons, including:

- The International Covenant on Economic, Social and Cultural Rights (Bangladesh acceded on 5 October 1998);

- The International Covenant on Civil and Political Rights (Bangladesh acceded on 6 September 2000);

- The Convention on Elimination of All Forms of Discrimination against Women (Bangladesh acceded on 6 November 1984); and

- The Convention on Rights of the Child (Bangladesh ratified on 3 August 1990).

Further, although non-binding, Bangladesh is bound to respect the UN Guiding Principles on Internal Displacement as they reflect and are consistent with international human rights and humanitarian law.

The UN Guiding Principles on Internal Displacement define an "internally displaced person" as "persons or groups of persons who have been forced or obliged to flee or to leave their homes or places of habitual residence, in particular as a result of or in order to avoid the effects of.....natural or human-made disasters, and who have not crossed an internationally recognized state border." Thus, the majority of persons displaced by the effects of climate change will be internally displaced

10

persons for the purposes of the Guiding Principles. Of particular relevance to climate displaced persons, the Guiding Principles provide:

- Internally displaced persons shall enjoy, in full equality, the same rights and freedoms under international and domestic law as do other persons in their country."
- They shall not be discriminated against in the enjoyment of any rights and freedoms on the ground that they are internally displaced.
- National authorities have the primary duty and responsibility to provide protection and humanitarian assistance to internally displaced persons within their jurisdiction.
- Internally displaced persons have the right to request and to receive protection and humanitarian assistance from these authorities. They shall not be persecuted or punished for making such a request.
- These Principles shall be applied without discrimination of any kind, such as race, color, sex, language, religion or belief, political or other opinion, national, ethnic or social origin, legal or social status, age, disability, property, birth, or on any other similar criteria.
- Certain internally displaced persons, such as children, especially unaccompanied minors, expectant mothers, mothers with young children, female heads of household, persons with disabilities and elderly persons, shall be entitled to protection and assistance required by their condition and to treatment which takes into account their special needs.
- All internally displaced persons have the right to an adequate standard of living.
- At the minimum, regardless of the circumstances, and without discrimination, competent authorities shall provide internally displaced persons with and ensure safe access to:
  (a) Essential food and potable water;
  (b) Basic shelter and housing;
  (c) Appropriate clothing; and
  (d) Essential medical services and sanitation.

Existing Policies and Laws in Bangladesh Relevant to Climate Change and Climate Displacement[10]

Bangladesh has adopted a number of laws, policies, strategies and institutional frameworks relevant to climate displacement. This chapter assesses the following 22 Policies, Plans, Projects, Acts, Standing Orders, Strategies and Programs of Action relevant to climate displacement in Bangladesh:

1. The National Adaptation Program of Action (2005)
2. The Bangladesh Climate Change Strategy and Action Plan (2009)
3. The National Plan for Disaster Management
4. The *Ashrayan* Project
5. The Disaster Management Act (2012)
6. The Standing Orders on Disaster (2010)
7. The Perspective Plan of Bangladesh
8. The National Strategy for Accelerated Poverty Reduction (2005)
9. Bangladesh Sixth Five Year Plan
10. The Bangladesh Country Investment Plan
11. National Agriculture Policy (2013)
12. The National Forestry Policy (1994)
13. The National Water Policy (1999)
14. The National Food Policy (2006)
15. The Coastal Zone Policy (2005)
16. The Coastal Development Strategy (2006)
17. The Environment Policy (1992)
18. The Bangladesh Environment Conservation Act (1995)
19. The National Housing Policy (2008)
20. The National Urban Sector Policy (2010)
21. The National Land Use Policy (2001)
22. The National Rural Development Policy (2001)

1. The National Adaptation Program of Action (2005): The National Adaptation Programs of Action are one of the types of reporting envisaged by the United Nations Framework Convention on Climate Change (UNFCCC). They are prepared by Least Developed Countries (LDC) to describe the country's perception of its most "urgent and immediate needs to adapt to climate change". The Least Developed Country Fund (LDCF) was established to finance the preparation of NAPAs and to implement the projects that they propose.

The Bangladesh National Adaptation Program of Action (NAPA, 2005) identifies 15 priority activities, including general awareness raising, capacity building, and project implementation in vulnerable regions, with a special focus on agriculture and water resources.

The first priority project of the NAPA has received financial support from the LDCF and is being implemented by the Ministry of Environment and Forest. The NAPA, produced in partnership with other stakeholders, highlights the main adverse effects of climate change and identifies adaptation needs.

The general objectives of the NAPA are:

• To make information about climate change impacts and adaptation available to decision makers;

• To incorporate potential adaptation measures into overall development planning process;

• To make development resilient to climate change; and

• To promote the sustainable development of Bangladesh.

The NAPA treats migration as an undesirable outcome of climate change. NAPA Priority Project No. 11 "Promoting Adaptation to Coastal Crop Agriculture to Combat Stalinization", mentions long term outcomes including community adaptation to flood, tidal surge and sea level rise. However, one of the stated goals of Project No. 11 is that "affected communities would not migrate to cities for job and livelihood" and the "social consequences of mass scale migration to cities would to some extent be halted". NAPA Project No. 12 "Adaptation to Agriculture Systems in Areas Prone to Enhanced Flash Flooding– North East and Central Region", states that desired long-term outcomes include "people might get a means to continue with farming, instead of migrating to cities after the flood. This would to some extent reduce social problems of migration of the distressed community to cities".

Migration can be a legitimate adaptation response to climate change and it is important that this negative portrayal of migration is updated and removed from these policy documents.

2. The Bangladesh Climate Change Strategy and Action Plan (2009): The main climate change strategic framework is the Bangladesh Climate Change Strategy and Action Plan (BCCSAP), published in 2008 and updated in 2009. The BCCSAP is a 10-year program (2009-2018) designed to build the capacity and resilience of the country to meet the challenge of climate change. The BCCSAP is designed as a 'living document' to continue to implement the nation's climate change adaptation and mitigation programs, as well as to deepen understanding of the phenomenon. It lists 44 different programs and 145 actions for implementation.

In the first five-year period (2009-13), the program was designed to comprise six pillars:

i) Food security, social protection and health;

ii) Comprehensive disaster management;

iii) Infrastructure;

iv) Research and knowledge management;

v) Mitigation and low carbon development; and

vi) Capacity building and institutional strengthening.

The Bangladesh Climate Change Strategy Action Plan makes clear reference to adaptation, mitigation, research and development, capacity building, institutional development, mainstreaming, disaster management and knowledge management. However, there is no mechanism for ensuring the implementation of these activities, as there is no implementation strategy.

In terms of addressing climate change migration and displacement, the BCCSAP states that "migration must be considered as a valid option of the country. Preparations in the meantime will be made to convert this population into trained and to be useful citizens for any country", suggesting the aspiration of displaced persons is to become 'useful citizens' by moving abroad. Whilst it is initially positive that migration is viewed as an effective and positive response to the effects of climate change, this broadly reflects the narrow policy of Bangladesh in that people displaced by climate change should migrate internationally, rather than within Bangladesh.

In the Research and Knowledge Management section of the BCCSAP, the Plan specifically requests that the Government establish a mechanism for the "monitoring of climate change related internal and external migration and rehabilitation".

3. The National Plan for Disaster Management (2010-2015): Bangladesh has a number of institutional structures to achieve technical monitoring, capacity building, preparedness and response. The National Plan for Disaster Management (NPDM) 2010-2015 is an outcome of the national and international commitments of the Government of Bangladesh.

The key targets, actions and outcomes to be achieved by 2015 under the National Plan for Disaster Management are organized under seven strategic goals:

- Professionalizing the disaster management system;
- Mainstreaming risk reduction;
- Strengthening institutional mechanisms;
- Empowering at risk communities;
- Expanding risk reduction programming;
- Strengthening emergency response systems; and
- Developing and strengthening networks.

The priority areas of focus for the NPDM are:

i) Articulate the long-term strategic focus of disaster management in Bangladesh;

ii) Demonstrate a commitment to addressing key issues including risk reduction, capacity-building, information management, climate change adaptation, livelihood security, issues of gender and the socially disadvantaged;

iii) Show the relationship between the vision of the Government, key result areas, goals and strategies and to align priorities and strategies with international and national drivers for change;

iv) Guide disaster management and risk reduction in the development and delivery of guidelines and programs;

v) Illustrate to other Ministries, NGOs, civil society and the private sector how their work can contribute to the achievements of the strategic goals and government vision on disaster management.

4. The Ashrayan **Project:** The objective of the *Ashrayan* project is to settle landless and homeless families (particularly those who are landless as a result of tropical cyclones, river erosion and floods) on *Khas* land and to provide those families with living accommodation and deeds of title jointly in the name of the husband and wife.

The *Ashrayan-1* Project was implemented from 1997 to 2010 and resettled and rehabilitated 108,646 families. *Ashrayan-II* is designed for implementation from 2010 to 2014 and intends to resettle and rehabilitate a further 50,000 families. *Ashrayan-II* has very recently been extended to 30 June, 2017.

The *Ashrayan* project aims to create dynamic villages and stimulate the socio-economic development of the people resettled by the project. The project is funded by the Prime Minister's Office and the physical implementation of the project is undertaken by the Armed Forces Division (AFD), different Government agencies, various Governmental Departments and District & *Upazila* Administrations. An

Ashrayan Project Central Advisory Council was created to ensure guidance and advice from appropriate authorities for the smooth implementation of the project.

The project has been working directly with civil and human rights issues in the target populations. Houses for persons from ethnic minorities are designed to maintain the culture and heritage of the resettled populations. Under the revised Development Project Proposal (DPP), multi-storey buildings will be constructed in City Corporation and Municipality areas at the District and *Upazila* levels to rehabilitate landless persons.

5. *The Disaster Management Act (2012)*: The Disaster Management Act (2012) has created mandatory obligations and responsibilities for Ministries, committees and appointments to ensure transparency in the overall disaster management system.

The objectives of the Disaster Management Act (2012) are:

- Substantial reduction of the overall risks of disasters to an acceptable level with appropriate risk reduction interventions;
- Effective implementation of post disaster rehabilitation and recovery measures;
- Emergency humanitarian assistance to the most vulnerable communities;
- Strengthening institutional capacity for the effective coordination of disaster management involving Government and NGOs; and
- Establishing a disaster management system capable of dealing with all existing hazards.

The Disaster Management Act is intended to promote a comprehensive disaster management program including an all-hazard, all-risk, all-sector approach where risk reduction, as a core element of disaster management, has equal emphasis with emergency response management and a greater focus on equitable and sustainable development.

6. *The Standing Orders on Disaster (2010)*: The Ministry of Disaster Management and Relief (MoDMR) launched an extensive consultation in order to finalize the updated version of the Standing Orders on Disaster (SOD). The SOD describe the roles and responsibilities of citizens, public representatives, NGOs, Ministries and other organizations in disaster risk reduction and emergency management. The SOD also established the necessary actions required to implement Bangladesh's disaster management model.

The SOD are intended to clarify the duties and responsibilities of all concerned people related to disaster management. All Ministries, Divisions, Departments and

Agencies prepare their own Action Plans in respect of their responsibilities under the Standing Orders.

The National Disaster Management Committee and Inter-Ministerial Disaster Management Coordination Committee ensure the coordination of disaster related activities at the National level. Coordination at District, *Upazila* and *Union* levels is ensured by the District Disaster Management Committee (DDMC), *Upazila* Disaster Management Committee (UZDMC) and Union Disaster Management Committee (UDMC). The Department of Disaster Management facilitates this process and renders necessary assistance.

7. The Perspective Plan of Bangladesh (2010-2021): The Perspective Plan of Bangladesh (2010-2021): *"Making Vision 2021: A Reality"* is a strategic articulation of the development vision, mission, goals and objectives to make Bangladesh a middle-income country.

A National Strategy plan was also developed to supplement this Perspective Plan. The National Strategy plan intends to reduce poverty from 40 percent to 15 percent by 2021. It specifies the key milestones along the way and highlights major intentions around strategic architecture, resources, competencies, and capacities.

The Perspective Plan has 8 development priorities which include mitigating the impacts of climate change. It has identified 18 management strategies to make Bangladesh resilient to the adverse impacts of climate change. The major management strategies are listed below:

• Stop environmental degradation through human activities through awareness-raising and, if necessary, recourse to legal means;

• Contain population growth, which is crucial for shaping the future of the country, and certainly in the context of climate change management;

• Best utilization of the available land, arresting and reversing the land degradation process;

• Conservation and enhancement of the country's biodiversity;

• Managing and improving sanitation in both rural areas and towns and cities;

• To improve navigability and water discharge, and to reduce flood risks, a strategy of dredging and training of rivers in a planned and phased manner will be pursued;

• Afforestation, particularly in coastal areas;

• Promotion of crop diversification;

- Integrated coastal zone management;
- Communities, particularly those to be affected most by extreme climatic events, will be the focus for capacity-building and mobilization;
- Resource mobilization from internal and external sources;
- All necessary steps will be taken to utilize nationally and internationally mobilized funds properly and effectively;
- Regional cooperation will be pursued for more effective flood and drought management as well as for basin-wide trans-boundary river management; and
- Climate change management through regional cooperation.

8. *The National Strategy for Accelerated Poverty Reduction (2005)*: The National Strategy for Accelerated Poverty Reduction: "Unlocking the Potential" was launched in October 2005. The Strategy lists eight specific avenues to achieve accelerated poverty reduction:

- Supportive macro-economic;
- Choice of critical sectors to maximize pro-poor benefits with special emphasis on rural, agricultural, informal and Small and Medium Enterprise (SME) sectors;
- Safety net measures to protect the poor;
- Human development through education;
- Health and sanitation;
- Participation and empowerment;
- Promotion of good governance;
- Improved service delivery; and
- Caring for environment.

The Poverty Reduction Strategy Paper II (2009-10) put emphasis on agricultural development considering the impacts of climate change induced natural disasters risk.

9. *The Bangladesh Sixth Five-Year Plan (2011-2015)*: According to the Ministry of Planning, climate change is not only an environmental issue, but also a challenge that could destabilize the economy. The Sixth Five Year Plan provides a detailed list of focus areas, including:

- Food security;
- Social protection;
- Health;

- Disaster management; and
- Infrastructure.

The Sixth Five Year Plan is based on the foundations of the National Adaptation Program of Action (NAPA) and the Bangladesh Climate Change Strategy and Action Plan (BCCSAP). The Sixth Year Plan also notes that the Government may need to strengthen existing institutions and may need to create and develop new institutions to respond effectively to the challenge of climate change.

10. The Bangladesh Country Investment Plan (2011-2015): The Bangladesh Country Investment Plan for Agriculture, Food Security and Nutrition was developed to support the implementation of the National Strategy for Accelerated Poverty Reduction. The Plan prioritizes 12 programs under 3 components, namely food availability, food access and food utilization. The first program focuses on integrated research and extension to develop sustainable responses to climate change.

11. National Agriculture Policy (2013): The Government of Bangladesh approved the "National Agriculture Policy 2013" with a focus on agriculture production, alleviating poverty through generating jobs, and highlighting food security. The agriculture policy has focused on high yielding varieties (HYV) of rice and other crops, fertilizer and mechanized irrigation over the decades. The overall objective of the National Agriculture Policy is to make the nation self-sufficient in food through increasing production of all crops including cereals and ensure a dependable food security system for all. The specific objectives of the National Agriculture Policy are to:

- Ensure a profitable and sustainable agricultural production system;
- Increase production and supplies of more nutritious food crops and thereby ensuring food security and improving nutritional status;
- Develop improved crop production technologies through research and training;
- Promote competitiveness through commercialization of agriculture land;
- Establish a self-reliant and sustainable agriculture adaptive to climate change and responsive to farmer's needs;
- Develop marketing system to ensure fair prices of agricultural commodities;
- Take necessary steps to ensure environmental protection as well as 'environment-friendly sustainable agriculture' through increased use of organic manure and strengthening of the Integrated Pest Management (IPM) programme; and

- Take appropriate steps to develop an efficient irrigation system and encourage farmers in providing supplementary irrigation during drought with a view to increasing cropping intensity and yield.

The National Agriculture Policy, 2013 has considered climate change issues and suggests promoting adaptation to climate change to reduce risk and vulnerability. Further, the Program Support Unit of the Ministry of Agriculture has prepared a climate management plan for the sector, which aims to provide a number of recommendations for climate change management planning within agriculture and farming systems for building a greater robustness and resilience to climate change factors and hazards.

The National Agriculture Policy states that the "Government of Bangladesh will have a contingency plan for taking up emergency agricultural rehabilitation programs (ARP) to recover from the crop losses due to any natural disaster at both the farmers' and national levels and this timeline will be on a short to long term basis. Also, the Government will take all kinds of cooperation from NGOs, entrepreneurs, private research and social service institutes for support disaster affected farmers." According to the National Agriculture Policy, the Government will develop "Agriculture disaster Recovery Grant and Crop Insurance" to support farmers affected by climate change induced disasters.

12. The National Forestry Policy (1994): The National Forestry Policy (NFP, 1994) recognizes the importance of biodiversity for environmental sustenance. The major objectives of the NFOP are:

- 20 percent of the total land area will be brought under afforestation programs;
- Bio-diversity of the existing degraded forests will be enriched by conserving the remaining natural habitats of birds and animals;
- The agricultural sector will be strengthened by extending assistance to the sectors related to forest development, especially by conserving land and water resources;
- Various international efforts and Government ratified agreements relating to global warming, desertification, control of trade in wild birds and animals will be implemented; and
- The illegal occupation of the forest lands, free felling and hunting of wild animals will be prevented, with the cooperation of local people.

13. The National Water Policy (1999): The National Water Policy (1999) provided the first comprehensive short, medium and long-term perspectives for managing water resources in Bangladesh. The Policy focuses on the importance of water for fisheries and wildlife, water for the environment and for preservation of wetlands.

The National Water Policy (1999) has 16 components, which describes policy measures to be undertaken to achieve the above objectives. These policy measures include:

- River basin management;
- Public and private involvement;
- Planning and management of water resources;
- Water rights and allocation;
- Public water investment;
- Water supply and sanitation;
- Water and agriculture;
- Water and industry;
- Water, fisheries and wildlife;
- Water and navigation;
- Water hydropower and recreation;
- Water for environment;
- Water for preservation of *haors*, boars, and beels;
- Economic and financial management;
- Research and information management; and
- Stakeholder participation.

14. The National Food Policy (2006): The goal of the National Food Policy (2006) is to ensure a dependable food security system for all people of the country at all times. The Policy clarifies three basic concepts:

i) Ensure adequate and stable supply of safe and nutritious food;
ii) Enhance purchasing power of the people for increased food accessibility; and
iii) Ensure adequate nutrition for all (especially women and children).

Given these basic concepts, the major objectives of the national food policy, which aims at ensuring dependable food security for all, are the following:

- Adequate and stable supply of safe and nutritious food at affordable   prices;

- Increased physical, social and economic access and purchasing power of all people; and
- Adequate nutrition for all individuals, especially children and women.

15. The Coastal Zone Policy (2005): The Coastal Zone Policy (CZP) recognizes the importance of ecosystem and biodiversity conservation. The Policy states that the coastal development process aims to meet, on an overall basis, the national goal for economic growth, poverty reduction and social development. The Policy also aims to abide by the Code of Conduct for responsible fisheries, the Code of Conduct for responsible mangrove management and other international conventions and treaties including the targets of the Millennium Development Goals.

The stated goal of the Policy for the Integrated Coastal Zone Management (ICZM) is "to create conditions, in which the reduction of poverty, development of sustainable livelihoods and the integration of the coastal zone into national processes can take place".

Section 4.2 of the Policy focuses on basic needs and opportunities for livelihoods. In this section it is mentioned that to meet basic needs of the coastal people and enhance livelihood opportunities, the Government policy should be as follows:

- Alleviation of poverty through creation of job opportunities and finding options for diversified livelihoods would be the major principles of all economic activities. Economic opportunities based on local resources will be explored to enhance income of the people;
- The Private sector and NGOs will be encouraged to implement activities for the poor people;
- Special measures will be taken during periods of disaster;
- *Khas* land will be distributed among the landless and a more transparent process of land settlement will be ensured; and
- An effective program for land reclamation will be developed.

16. The Coastal Development Strategy (2006): The Coastal Development Strategy (2006) links the Coastal Zone Policy (CZP) with development programs and interventions. The objectives of the Strategy are: "to select strategic priorities and actions in implementation of the CZP with emphasis on the creation of the institutional environment that will enable the Government of Bangladesh to embark on a continuous and structured process of prioritization, development and implementation of concerted interventions for the development of the coastal zone".

The Coastal Development Strategy (2006) describes the priorities and targets based on the Coastal Zone Policy (CZP) objectives, the problems and issues in the coastal zone and the available resources. The Strategy represents a departure from 'business as usual' in the management of the coastal zone towards utilizing its potential. It describes 'governance' of the coastal zone. The Strategy takes into account emerging trends, including increasing urbanization, changing patterns of land use, declining land and water resources, unemployment and visible climate change impacts. However, there is no effective land distribution strategy for those whose land is lost due to erosion.

17. The Environment Policy (1992): The Environmental Policy built upon the 1992 United Nations Conference on Environment and Development (The Rio Conference) and acknowledged that sustained development of the country was based on the well-being of the environment and ecosystems as they provide the services necessary for ensuring progress.

The objectives of the Environment Policy are:

• Maintain ecological balance and overall development through protection and improvement of the environment;

• Protect the country against natural disasters;

• Identify and regulate activities which pollute and degrade the environment;

• Ensure environmentally sound development in all sectors; and

• Ensure sustainable, long term and environmentally sound use of all national resources.

The Environment Policy recognized the need for a better and more comprehensive approach to address climate change and environment issues. Policies towards realization of the overall objectives of the Environment Policy are described in 15 sectors:

i)   Agriculture;

ii)   Industry;

iii) Health and Sanitation;

iv)   Energy and Fuel;

v)   Water Development, Flood Control and Irrigation;

vi)   Land;

vii) Forest, Wildlife and Bio-diversity;

viii) Fisheries and Livestock;

ix)   Food;

x)   Coastal and Marine Environment;

xi)   Transport and Communication;

xii) Housing and Urbanization;

xiii) Population;

xiv) Education and Public Awareness; and

xv) Science, Technology and Research.

18. The Bangladesh Environment Conservation Act (1995): The Bangladesh Environment Conservation Act (1995) provides for conservation and improvement of environmental standards and for controlling and mitigating environmental pollution. However, the Act provides very few substantive obligations relating to environmental clearance from the Department of Environment and any person affected or likely to be affected by such activities can apply to the Director General seeking remedy of environmental pollution or degradation. The major limitations of the Act are its silences on the standards, parameters, emission levels and management elements based on which the environmental clearance should have been applied and obtained.

The Act empowers the Government to declare an area to be an "ecologically critical area" if its eco-system appears to be under serious threats of degradation and vulnerable to climate change. The Ministry of Environment and Forest has already declared seven areas as critical. These include the *Sundarbans*, Cox's Bazar-*Teknaf* sea beach, St. Martin's Island, *Shonadia* Island, *Hakaluki Haor*, *Tanguar Haor*, and *Marzat Oxbow* Lake. In these areas, a ban has been imposed on some activities that include felling or extracting trees; hunting and poaching of wild animals; catching or collection of snails, coral, turtles, and other creatures; any activities that may threaten the habitat of flora and fauna; activities likely to destroy or alter the natural characteristics of the soil and water; establishment of industries that may pollute soil, water, air and/or create noise pollution and any other activity that may be harmful for fish and aquatic life.

19. The National Housing Policy (2008): The objectives of the National Housing Policy are to "make housing accessible to all strata of society.... the high priority target groups will be the disadvantaged, the destitute and the shelter-less poor; and to develop effective strategies for reducing the need to seek shelter through formation of slums.... to relocate them in suitable places".

The Policy states that the Government recognizes the difficult situation in which the poor live in slums and squatter settlements. The main objectives of the National Housing Policy are:

- Ensuring housing for all with particular emphasis on the disadvantaged, destitute, the shelter less poor and the low and middle-income groups of people;
- Make available suitable land for housing at affordable price;
- Developing mechanisms to discharge formation of slums and squatter settlements, unauthorized constructions and encroachments;
- Mobilization of resources for housing through personal savings and financial institutions;
- Developing institutional and legal framework for facilitating housing; and
- Providing encouragement to universities, research institutions and research centers for housing oriented researches.

The strategy of the Government for implementing the National Housing Policy is to act as a promoter and facilitator and, to a limited extent, as a provider.

The salient features of the housing strategy envisaged in the National Housing Policy are:

i) Housing will be given due priority in national development plans;

ii) The role of the Government in housing will be to supply serviced land at reasonable prices and to help create and promote housing financing   institutions;

iii) Efforts will be made to increase affordability for the disadvantaged and the low income groups through providing credit for income generation;

iv) Improvement and rehabilitation of the existing housing stock will be given priority by the Government alongside new housing; and

v) Ensuring the conservation of the natural environment and preservation of cultural heritage in new housing projects.

The Policy specifically provides in paragraph 5.8.9 that "rehabilitation for river erosion and other natural disaster affected communities should be ensured in the village land bank". Housing reconstruction and rehabilitation in disaster prone areas is one of the major components of the National Housing Policy.

Paragraph 5.10.2 states that "necessary action should be taken urgently for reconstruction and re-building of houses damaged by cyclones, floods and other natural disasters. Special rehabilitation programs including easy terms for housing loans should be ensured for communities in disaster prone areas".

20. *The National Urban Sector Policy (2010)*: The National Urban Sector Policy (2010) envisions strengthening the beneficial aspects of urbanization and at the same time effectively dealing with its negative consequences so as to achieve sustainable urbanization, keeping in view the multi-dimensional nature of the urbanization process. The policy is designed to be gender sensitive and sensitive to the needs of children, the aged and the disadvantaged.

The key components of the policy are:

i)   Patterns and process of urbanization;
ii)   Local urban planning;
iii)   Local economic development and employment;
iv)   Urban local finance and resource mobilization;
v)   Urban land management;
vi)   Urban housing;
vii)   Urban poverty and slum improvement;
viii) Urban environmental management;
ix)   Infrastructure and services;
x)   Social structure;
xi)   Rural-urban linkage;
xii)   Urban governance; and
xiii) Urban Research, Training and Information.

The Policy recognizes that urban areas will form a network of distribution where each centre will fall into a hierarchy. This policy also recognizes that rural to urban migration plays a key role in urbanization and that it has both positive and negative consequences. To achieve balanced urbanization rural-urban migration must be properly guided to avoid over concentration of population in one or few cities.

The Policy has a special focus on Urban Land Management (paragraph 5.5). The Policy emphasizes that the Government must exert some degree of control over the use and development of urban land based policies and regulations. A range of urban planning tools including land use planning, transportation planning and management, site planning, subdivision regulations and building regulations can be applied to minimize environmental impacts of urban development activities.

In the Urban Land Management section (paragraph 5.5.7), the Policy emphasizes:

i) Reforming land transfer laws to counter trends towards land accumulation;

ii) Implementation of land-banking and land-pooling programs that allow the government to increase its pool of land which can be exchanged for low-cost housing sites in the city;

iii) Undertaking land readjustment projects that include low-cost land and housing sites;

iv) Allocating *Khas* land/acquired land for housing the poor; and

v) Allocating a reasonable proportion of land in urban places for housing the poor.

The policy also highlights issues related to land development (paragraph 5.5.10) where it is stated that the Government can intervene in the land market either by developing the land itself or by facilitating the private sector to carry out land development activities and also take up special schemes to develop land for housing low-income groups and the poor.

The National Urban Sector Policy makes provision for in-situ upgrading and improvement of slums, resettlement of slum dwellers and seeks to ensure tenure security for the urban poor. The Policy states that there should not be any eviction of slum dwellers without proper rehabilitation. The Policy also states that master plans should designate areas for slum rehabilitation and that the Government should provide the urban poor with access to infrastructure and services to all inhabitants of slum/informal settlements. The Policy also makes provision for the allocation of land and finances for slum improvement programs in all urban areas.

21. The National Land Use Policy (2001): The National Land Use Policy (2001) highlights the need, the importance and the modalities of land zoning for integrated planning and management of land resources of the country. The Policy emphasizes the distribution of *Khas* lands among landless people in Bangladesh. *Khas* land is Government owned land and applies to agricultural land, non-agricultural land and water bodies. However, these programs have met with mixed success due to vested interests illegally occupying *Khas* land, a lack of political will, the inefficiencies in the way local and national administration is organized and the absence of an updated, systematic and universally accepted source of information on land resource availability and land rights.

It is essential that the Government take steps to implement an effective, transparent and just program for the distribution of *Khas* land to landless persons - including climate displaced persons. These programs should be rights-based, they

27

should involve the participation of affected communities in their design and the ability to review adverse decisions should be clear and accessible. It is important that decisions about the distribution of *Khas* land are made on the basis of genuine need, rather than political or personal considerations. Civil Society representatives should also be part of the decision-making panels for *Khas* land distribution. Furthermore, training should be provided to decision makers on climate displacement in Bangladesh and the need to ensure rights-based durable solutions for climate displaced persons.

The Policy also mentions the need to formulate a Zoning Law and Village Improvement Act for materializing the identified land zoning area. The Policy highlights the need for land zoning for the coastal area of Bangladesh. It also describes the need for definite guidelines and raises the possibility of undertaking coastal land zoning through an inter-ministerial task force.

22. The National Rural Development Policy (2001): The National Rural Development Policy consists of 30 programs under seven sections, these are:

- People's Participation;
- Poverty Alleviation;
- Rural Infrastructure Development;
- Agro-based Rural Economy;
- Education for Rural Areas;
- Rural Health Service and Nutrition Development;
- Rural Population Control;
- Development of Rural Housing;
- Land Use and Development;
- Rural Industries Development;
- Rural Capital Flow and Financing;
- Empowerment of Rural Women;
- Rural Child and Youth Development;
- Development of Disadvantaged Rural People;
- Area Specific Special Development Programs;
- Employment Generation for Self-Reliance;
- Creation of Skilled Manpower in Rural Areas;
- Cooperatives for Rural Development;

- Rural Environment;
- Dispute Settlement/ *Salish* System;
- Law and Order;
- Culture and Heritage;
- Games and Sports;
- Power and Fuel Energy;
- Research and Training;
- Information Dissemination and Data Base;
- Awards for Contributions to Rural Development;
- Contribution by NGOs and Other Actors;
- Support to Elderly People; and
- Regional and International Cooperation.

The Policy emphasizes the integration of all activities in rural development with a view to alleviating poverty; improving the quality of life of women and the poor and the economic development of landless and marginal farmers. Relevant provisions of the Act include:

Paragraph 5.8 of the Act relates to the "Development of Rural Housing" and Article 6 states:

"Families, who become landless, displaced or shelter-less due to river erosion, will be provided with shelter within a short time frame on a priority basis and will be rehabilitated in the nearest Government *Ashrayan/ Adarsha Gram* project area". Paragraph 5.9 relates to Land Use and Development and Article 4 states:

"Giving priority to the use of land for rural poverty alleviation will be continued and ensured in the allocation, distribution and leasing out of *Khas* land and Government water bodies".

Report on Climate Displacement in Bangladesh: *The Need for Urgent Housing, Land and Property Rights Solutions*

Climate Displacement in Bangladesh: *The Need for Urgent Housing, Land and Property Rights Solutions*, May 2012 is a 44-page report in English with a translation in Bengali, published by Displacement Solutions and YPSA. This report comprehensively examines the scope and causes of climate displacement across Bangladesh. Drawing on extensive fieldwork, the report highlights that climate displacement is not just a phenomenon to be addressed at some point in the future, it is a crisis that is unfolding across Bangladesh now. Sea-level rise and tropical cyclones in coastal areas, as well as flooding and riverbank erosion in mainland areas, are already resulting in the loss of homes, land and property and leading to mass displacement. Further, all of the natural hazards that are causing displacement are expected to increase in both frequency and intensity as a result of climate change – almost inevitably leading to the displacement of many millions more across Bangladesh.

This report is designed to develop awareness and deepen knowledge of this crucial issue as well as to propose concrete, practical recommendations that can be utilized by the Government of Bangladesh, civil society actors, climate affected communities themselves, academics, development practitioners, the regional and international communities and other relevant stakeholders.

This report comprehensively examines current and future causes of climate displacement in Bangladesh. The report also examines existing and proposed Government and civil society policies and programs intended to provide solutions to climate displacement. The report highlights a number of protection gaps in the response of both the Government of Bangladesh and the international community to the plight of climate displaced persons. The report emphasizes that rights-based solutions, in particular, housing, land and property rights solutions must be utilized as the basis for solving this crisis.

The report concludes by proposing a number of concrete recommendations that could be utilized to provide solutions to climate displacement.

Study on Land Availability for Climate Displaced Communities in Bangladesh

DS and YPSA jointly prepared the report of Land Availability for Climate Displaced Communities in Bangladesh after intensive study in Bangladesh (this report published as chapter-4 of *Land Solutions to Climate Displacement*, edited by Scott Leckie in 2014). The objectives of the study were categorization of each of the various land categories in Bangladesh, including both public and private land; analysis of land categories suitable for the resettlement of climate displaced persons in Bangladesh; estimate of the land resources required to resettle the entire climate displaced population in Bangladesh; and identification of specific land parcels which could be acquired and accessed by civil society groups and climate displaced communities an utilized as possible resettlement sites.

This report examines whether there is indeed enough land available in Bangladesh to find new land plots for the current and future millions of climate displaced persons within Bangladesh. The report concludes with the following major recommendations;

- To ensure the successful planned rehabilitation and resettlement of climate displaced people, Bangladesh must be adequately prepared so that the vast majority of those displaced can be supported with adequate resettlement and rehabilitation schemes;

- Government owned *khas* land should play a key role in providing land solutions to climate displacement in Bangladesh;

- Suitable *khas* land should be free from illegal occupancy for undertaking resettlement scheme for climate displaced persons; and

- Long term policy advocacy and campaign strategies will be required to free illegally occupied *khas* land and facilitate access of climate displaced people to those lands.

Study on Land Acquisition for Climate Displaced Communities in Bangladesh

DS and YPSA also prepared the report of Land Acquisition for Climate Displaced Communities in Bangladesh after intensive review of relevant laws, policies and practices in Bangladesh (this report later published as chapter-5 of *Land Solutions to Climate Displacement*, edited by Scott Leckie in 2014). The major objective of this Study was to explore in detail the many legal, social, historical, political, economic and other factors involved in the land acquisition process in Bangladesh.

This Study Report examined in detailed how land across Bangladesh can most easily, affordably and fairly be acquired and accessed by civil society groups and climate displaced communities in Bangladesh. The study also assessed the possibility of climate displaced persons, who have lost their land, accessing land through private or public donation, including by government officials, private individuals or corporations. The report concluded with recommendation of necessary improvements - the distribution of agricultural *khas* land and amendments - the distribution of non- agricultural *khas* land under the 1990 Land Management Manual and the Char Development and Settlement Project (CDSP) could both provide land solutions to climate displacement in Bangladesh. Overall for the rehabilitation of climate displaced persons, there should be a comprehensive rehabilitation policy in place.

Study on The Viability of the Chittagong Hill Tracts as a Destination for Climate Displaced Communities in Bangladesh

DS and YPSA also prepared the report of The Viability of the Chittagong Hill Tracts as a Destination for Climate Displaced Communities in Bangladesh after intensive review of relevant laws, policies and present practical scenarios in Bangladesh (this report later published as chapter-6 of *Land Solutions to Climate Displacement*, edited by Scott Leckie in 2014). The major objectives of this study were to examine the political and historical sensitivities of the CHT and indicate areas of focus for future activities in the CHT in this regard as well as to observe the potential environmental impacts of resettling large numbers of climate displaced persons in the CHT. The study examined in detailed to draw recommendations on the political, social, economic, environmental and other types of viability for the CHT as a possible permanent destination for climate displaced communities wishing to resettle there.

The report concluded with recommendation that due to social, political, economic and environmental factors - the Chittagong Hill Tracts are not a suitable destination for the majority of climate displaced persons. The CHT is one of the most disadvantaged and vulnerable regions in the country in terms of almost all major development indicators, including income, employment, poverty, health, water, environment and sanitation, education, women's employment, access to infrastructure and national building institutions, peace and inter community confidence. The geo-political situation of the CHT makes the area vulnerable, as it shares a border with Myanmar and with India. Besides, the 1997 Peace Accord made the following modifications of control over land law in the CHT, which is mentioned in clause 26 under section B of the 1997 Peace Accord: By amendment of the section 64 the following sub-sections shall be made—*Notwithstanding anything contained in any law for the time-being in force, no land within the boundaries of Hill District shall be given in settlement, purchased, sold and transferred including giving lease without prior approval of the Council.* This amendment of law indirectly stops the provision of land acquisition by outsiders in the CHT region.

Climate Displacement in Bangladesh: Stakeholders, Laws and Polices - Mapping The Existing Institutional Framework

This 182 pages mapping study demonstrates that there are a large number of Government and Non-Government stakeholders at the national, regional and international levels, as well as a large number of laws and policies that are directly or indirectly relevant to climate displacement in Bangladesh. However, despite this abundance of stakeholders and laws, at present they do not combine to create a coherent, comprehensive or effective institutional framework for responding to or planning for climate displacement in Bangladesh.

The mapping study seeks to identify and clarify the existing institutional framework as it relates to climate displacement in Bangladesh. The mapping study also recommends a number of steps that could be taken to improve the existing institutional framework, in order to better ensure the human rights and specifically the housing, land and property rights of all climate displaced persons in Bangladesh.

To this end, the mapping study identifies 168 institutional and organizational stakeholders and 78 resource persons at the national, regional and international levels, including:

- 36 Government Ministries, Departments, Institutes and Authorities;
- 20 International Donors and Funding Organizations;
- 14 National Civil Society Organizations and Networks;
- 45 National NGOs;
- 23 International NGOs;
- 30 Academic Institutes, Research Centers; and
- 78 National Experts.

The mapping study also identifies and assesses 22 Laws, Policies, Strategies and Programs of Action relevant to climate displacement in Bangladesh.

The mapping study provides three practical diagrams, offering guidance on the institutional structure as it applies during natural disaster and climate related displacement, as well as the *Khas* land distribution process in Bangladesh for the rehabilitation of persons affected by climate change induced natural disasters:

1. The current institutional structure to respond to temporary displacement due to natural disasters or climate change in Bangladesh;

2. The current institutional structure for the relocation of persons displaced by natural disasters or climate change in Bangladesh;

3. The current structure of *Khas* land distribution to landless and homeless persons in Bangladesh.

The study concludes with seven recommendations for improving the existing institutional framework; especially how the institutional framework can better ensure the human rights of all climate displaced persons in Bangladesh:

1. A comprehensive review and analysis of existing national laws and policies related to climate displacement should be undertaken;

2. Rights-based national laws and policies on climate displacement should be designed and implemented;

3. A review and re-design of the existing institutional framework on climate displacement should be undertaken;

4. Domestic land solutions for climate displaced persons should be designed and implemented;

5. Coordination and knowledge sharing among international and regional donors on climate displacement should be increased;

6. A United Nations and NGO working group on climate displacement in Bangladesh should be created; and

7. Coordination and communication between Government and civil society on climate displacement should be improved.

This mapping study is designed to be used by Government officials and representatives in Bangladesh as well as all stakeholders at the national, regional and international levels who are working directly or indirectly on climate displacement in Bangladesh. It is also anticipated that this process of mapping and assessing the climate displacement institutional framework could be usefully replicated in other countries also affected by climate displacement.

Formulation of Peninsula Principles on Climate Displacement

To set the rules to assist governments to provide solutions for climate change induced displacement recorded countries, representatives from Australia, New Zealand, Alaska, Bangladesh (YPSA as partner of DS), Netherlands, Switzerland, UK, Germany, Egypt, Tunisia and the US got together in Red Hill, Victoria on August 2013. After months of preparatory work around the world, an eminent group of international lawyers, UN officials and climate change experts, from the above mentioned countries, have finally agreed on new global rules, that considered as "Peninsula Principles on Climate Displacement" (a name inspired by the Mornington Peninsula of Australia where the meeting took place) outlining the rights of people and communities who lose their homes, land and livelihoods due to the effects of climate change already underway in many – primarily Asia-Pacific -countries such as Bangladesh, the Maldives, Papua New Guinea, Solomon Islands, the US and elsewhere. In the meeting participants shared their backgrounds and expertise in International Law, migration, forced migration, environmental change and UN policy creation to strengthen and stand behind the Peninsula Principles as the first formal policy of its kind in the world. This new global rules published with the title of "The Peninsula Principles, *on climate displacement within states"* primarily in English and later with a translation in Bengali by Displacement Solutions and YPSA.

YPSA has contributed strongly for the formulation of first global policy by Displacement Solution on climate displacement 'Peninsula Principles on Climate displacement within State" in Australia.The Peninsula Principles are developed on the basis of current international law; several thousand interviews carried out over the past five years in heavily affected countries and were most recently influenced by comments received from the public at large who had access to the draft Principles on the internet.

These Peninsula Principles provide a comprehensive normative framework, based on principles of international law, human rights obligations and good practice, within which the rights of climate displaced persons can be addressed within a State and not cross-border climate displacement. This principles set out protection and assistance principles, consistent with the UN Guiding Principles on Internal Displacement, to be applied to climate displaced persons. Peninsula Principles consisted of 18 sub titles as follows;

➢ Principle-1: Scope and Purpose

- Principle-2: Definitions
- Principle-3: Non-discrimination, rights and freedoms
- Principle-4: Interpretation
- Principle-5: Prevention and avoidance
- Principle-6: Provision of adaptation assistance, protection and other measures
- Principle-7: National implementation measures
- Principle-8: International cooperation and assistance
- Principle-9: Climate displacement risk management
- Principle-10: Participation and consent
- Principle-11: Land Identification, Habitability and Use
- Principle-12: Loss and Damage in the context of displacement
- Principle-13: Institutional Frameworks to support and facilitate the provision of assistance and protection
- Principle-14: State assistance to those climate displaced persons experiencing displacement but who have not been relocated
- Principle-15: Housing and livelihood
- Principle-16: Remedies and compensation
- Principle-17: Framework for return
- Principle-18: Implementation and dissemination.

The key features of the Peninsula Principles are:

States shall not discriminate against climate displaced persons on the basis of their potential or actual displacement, and should take steps to repeal unjust or arbitrary laws and laws that otherwise discriminate against, or have a discriminatory effect on, climate displaced persons. Climate displaced persons shall enjoy, in full equality, the same rights and freedoms under international and domestic law as do other persons in their country, in particular housing, land and property rights.

States should provide adaptation assistance, protection and other measures to ensure that individuals, households and communities can remain in their lands or places of habitual residence for as long as possible in a manner fully consistent with their rights.

States should incorporate climate displacement prevention, assistance and protection provisions as set out in these Peninsula Principles into domestic law and policies, prioritizing the prevention of displacement. States should ensure that durable solutions to climate displacement are adequately addressed by legislation and other administrative measures.

States that are otherwise unable to adequately prevent and respond to climate displacement should accept appropriate assistance and support from other States and relevant international agencies, whether made individually or collectively.

With regard to climate displacement risk management, monitoring, and modeling, States, using a rights-based approach, should identify, design and implement risk management strategies, including risk reduction, risk transfer and risk sharing mechanisms, in relation to climate displacement as well as undertake systematic observation and monitoring of, and disaggregated data collection at the household, local, regional and national levels on, current and anticipated climate displacement.

To enable successful preparation and planning for climate displacement, States should ensure that priority consideration is given to requests from individuals, households and communities for relocation. And no relocation shall take place unless individuals, households and communities (both displaced and host) provide full and informed consent for such relocation.

Recognizing the importance of land in the resolution of climate displacement, States should identify, acquire and reserve sufficient, suitable, habitable and appropriate public and other land to provide viable and affordable land-based solutions to climate displacement. States should also plan for and develop relocation sites including new human settlements on land not at risk from the effects of climate change or other natural or human hazards and, in so planning, consider the safety and environmental integrity of the new site(s), and ensure that the rights of both those relocated and the communities that host them are upheld.

States should develop appropriate laws and policies for loss suffered and damage incurred in the context of climate displacement.

States should strengthen national capacities and capabilities to identify and address the protection and assistance needs of climate displaced persons through the establishment of effective institutional frameworks and the inclusion of climate displacement in National Adaptation Program of Action as appropriate.

States have the primary obligation to provide all necessary legal, economic, social and other forms of protection and assistance to those climate displaced persons experiencing displacement but who have not been relocated. Protection and assistance activities undertaken by States should be carried out in a manner that respects both the cultural sensitivities prevailing in the affected area and the principles of maintaining family and community cohesion.

States should respect, protect and fulfill the right to adequate housing of climate displaced persons experiencing displacement but who have not been relocated, which includes accessibility, affordability, habitability, security of tenure, cultural adequacy, suitability of location, and non-discriminatory access to basic services (for example, health and education).

Climate displaced persons experiencing displacement but who have not been relocated and whose rights have been violated shall have fair and equitable access to appropriate remedies and compensation.

States should allow climate displaced persons experiencing displacement to voluntarily return to their former homes, lands or places of habitual residence, and should facilitate their effective return in safety and with dignity, in circumstances where such homes, lands or places of habitual residence are habitable and where return does not pose significant risk to life or livelihood.

States, who have the primary obligation to ensure the full enjoyment of the rights of all climate displaced persons within their territory, should implement and disseminate these Peninsula Principles without delay and cooperate closely with inter-governmental organizations, non-government organizations, practitioners, civil society, and community based groups toward this end.

These Peninsula principles are guided by the Charter of the United Nations, and Reaffirming the Universal Declaration of Human Rights, the International Covenant on Economic, Social and Cultural Rights, the International Covenant on Civil and Political Rights as well as the Vienna Declaration and Program of Action. Assuming that the international community has humanitarian, social, cultural, financial and security interests in addressing the problem of climate displacement in a timely, coordinated and targeted manner and finally States will bear the primary responsibility for their citizens and others living within their territory, but recognizing that, for many States, addressing the issue of and responding to climate displacement presents financial, logistical, political, resource and other difficulties.

Advocacy and Capacity Building on the Rights of Climate Displaced Persons in Bangladesh

YPSA established an informal network with the likeminded local level organizations working in the climate change filed in the different climatic vulnerable areas of Bangladesh for successful advocacy and capacity building of local level stakeholders including government officials, civil society organizations, local government and representatives of climate displaced communities. These network organizations were involved during the different field study and organizing advocacy meeting in the respective area. YPSA has conducted capacity building event at Sub district and district level, as well as in the National level on the Rights of Climate Displaced Persons in Bangladesh. Different types of IEC materials (study reports, booklet, poster, folder with message and stickers) also developed and printed to accelerate the capacity building initiatives. These materials also utilized in all advocacy events and one to one session.

YPSA has also undertaken extensive advocacy with local, regional and national Government representatives, academics, media professionals, civil society organisations, lawyers associations, international NGOs and UN agencies in order to encourage the development effective legal and policy solutions to climate displacement in Bangladesh. The advocacy also included two national level Round Table Discussion on climate displacement with the title "Climate Displacement in Bangladesh: The Need for Urgent Housing, Land and Property Rights Solutions" with relevant stakeholders. Chairman of Parliamentary standing Committee for Ministry of Environment and Forest said as a chief guest in the Round Table that 'A comprehensive Climate Change Policy is now demand of time, which can cover all climate changes issues including the rights of climate displaced peoples of Bangladesh. I, myself, will take necessary initiative to fulfill this requirement'. To raise awareness of the situation of climate displacement in Bangladesh, the rights of climate displaced persons and the work and plans of the Bangladesh HLP Initiative, DS and YPSA met with representatives from UNDP, UNHCR, the Embassy of Sweden, the Embassy of Germany, Oxfam, GIZ, PKSF, AFD, IFAD, the Embassy of the Netherlands and the Delegation of the EU to Bangladesh.

The Bangladesh HLP Initiative provided orientation to the leading media representatives from print, online and electronic media for raising mass awareness on the rights of climate displaced persons. Journalists were asked to publish news /

documentary about the worst situation of climate displaced persons of Bangladesh. Journalists were agreed to report on the newly burning issue of climate displacement in their respective media seeking support to visit different hotspot creating climate displacement. Later on YPSA initiated large scale media campaign with print, electronic and online media journalist for raising awareness on climate displacement issues in Bangladesh. Before organizing and sending the journalist group to different hot spot, YPSA briefed the objective of this visit and sought support from media to act proactively to highlight the displaced people rights based on their sufferings. Till to date more than 20 news, articles published/ broadcasted in the many print, online and electronic media.

DS and YPSA have conducted several advocacy meetings with senior representatives from the rolling Awami League, the Jatiya Party and the Bangladesh Nationalist Party of Bangladesh Political Parties. All the party's leader morally agreed on multifaceted climate change effects on people's life and livelihood, national economy and national development as a whole. They also shared their experiences of distressed condition of displaced people, who lost all their belongings and most of the time these landless families are forced to live on the embankment or permanently switch to some other places. They also emphasized that proper steps needs to be taken to reduce displacements and side by side more rehabilitation activities need to be undertaken. They expressed their hope that proper monitoring of policy implementation and mass awareness on displacement issue will help these displaced communities to get justice. They also focused on mass awareness among vulnerable people about their rights and climate change impacts. During the meeting YPSA hand over them advocacy materials for better settlement of climate displaced peoples and expected to raise their voice on the issue in the Parliament house.

YPSA also doing advocacy with the lawyers association of Chittagong as part of Bangladesh HLP rights initiative. The objectives of this advocacy are – i) to raise awareness legal groups and Lawyers about the international and national legal issues related on climate displacement and ii) identify the duties and responsibilities of the legal groups and lawyers for ensuring housing, land and property rights relevant to climate displacement through forming a legal forum on climate displacement issue. YPSA also facilitated the creation of the "Lawyer's Initiative for Displacement Solutions" in the Chittagong Bar Association. This informal network of local lawyers in Bangladesh intends to undertake lobbying on effective legislative and policy change as well as public interest litigation under the Constitution of Bangladesh on the rights of climate displaced persons.

Coastal Kids Project

The Coastal Kids Project, as a sub component of Bangladesh HLP Initiative, is designed to enhance understanding of the effects of climate change through educating children of Australia, Bangladesh, Kiribati and Tuvalu (countries particularly threatened by climate change) about climate change issues and facilitating direct contacts among the children with the same age group of those countries. The main focus of Coastal Kids Project is to build strong bondage among the students (between the ages of 10 to 12) of Bangladesh, Australia, Kiribati and Tuvalu who are living in coastal areas for sharing knowledge and exchanging views through Skype discussions about climate change issues. The following activities successfully completed under this project.

- Signed MoU with School
- Provided orientation to student on climate change and displacement issue
- Organized Skype conversation between two schools of Bangladesh and Australia
- Organized Annual Coastal Kids writing competition and replied to pen-pal letters
- Organized debate competition for the schools on Climate Change issues
- Organized art competition on Climate Change issues
- Facilitated photo competition on Climate Change issues

In all the events coastal kids participated with full enthusiasm and enjoyed very much. All competition successfully ended followed by very nice prizes.

Development of Guidance Note of Access to *Khas* Land for Climate Displaced Persons in Bangladesh

After the exclusive interview with land officials, NGO official working with land issue and reviewing the existing *khas* (government owned) land related documents, YPSA developed, reviewed by Displacement Solutions (DS), the Guidance Note of "Access to *Khas* Land for Climate Displaced Persons in Bangladesh", which seeks to

41

clarify how climate displaced persons and their advocates can access and utilize these existing processes to access new parcels of land within Bangladesh. This Guidance Note is also intended to be useful for Government officials and representatives seeking to improve the current *khas* land distribution processes. Finally, it is intended that this Guidance Note will be useful for the regional and international communities – for the challenge of climate displacement in Bangladesh is not one to be faced alone but must be addressed with regional and international support and cooperation.

The Guidance Note briefly described the eleven stages of *khas* land distribution based on existing policy and manuals. This also includes the *khas* land distribution and Land settlement and titling activities for landless peoples under Char Development and Settlement Project. Before finalization of guidance note, it is thoroughly verified by two land experts working as Land official of Bangladesh Government *khas* land distribution process whether it is in line with the Bangladesh Government *Khas* land distribution Policy and strategy and finally they endorsed this guidance note of "Access to *khas* land for climate displaced persons in Bangladesh" that is in line with government policy and manuals.

The following ordinances, manuals, programmes and policies provide the basis for the distribution of agricultural *khas* land under the Land Management Manual:

- The 1984 Land Reform Ordinance
- The 1987 Land Reform Action Program
- The 1997 Agricultural *Khas* Land Management and Settlement Policy (and subsequent amendments)

Based on these laws and policies, the process for distributing *khas* land has been divided into the following 11 stages from identification of *khas* land to handover the *khas* land to applicant landless families as follows:

☐ The identification of *khas* land
☐ Publication of notice of all agricultural *khas* lands
☐ Preparation of *khas* land for distribution
☐ Applications for agricultural *khas* Land from landless families
☐ Selection of landless families for agricultural *khas* land
☐ Decision process for plot distribution to landless families
☐ Preparation of case file for settlement
☐ *Khas* land settlement
☐ Registration of distributed *khas* land

☐ Formal meeting for handover of the *khas* land settlement document

☐ Conditions of *khas* land possession

NGOs can also support land less people through facilitating the entire process of agriculture *khas* land distribution as mentioned in the Land Management Manual, 1990.

The Char Development and Settlement Project (CDSP) works to support families to build their lives on this new land since 1994, one of the aims of the CDSP is to improve security of tenure for persons living in *char* areas, through providing households with a secure land title. The land settlement process under the CDSP-IV follows the provisions of the Agricultural *Khas* land Management and Settlement Act and can be divided into the following 3 stages :

☐ Production of settlement map

☐ Consolidation of the map and information on landless households

☐ Issuance of the Official Land Title

This Guidance Note seeks to clarify how climate displaced persons and their advocates can access and utilize these existing processes to access new parcels of land within Bangladesh. This Guidance Note is also intended to be useful for Government officials and representatives seeking to improve the current *khas* land distribution processes. Finally, it is intended that this Guidance Note will be useful for the regional and international communities for the challenge of climate displacement in Bangladesh is not one to be faced alone but must be addressed with regional and international support and cooperation.

Five Key Actions for ensuring the rights of climate displaced persons

To strengthen the advocacy campaign on climate displacement in Bangladesh the Bangladesh HLP initiative, consultation with different stakeholders, developed the five key actions including reference to relevant domestic and international legal standards. This advocacy document identifies five key actions that could be undertaken by the Government of Bangladesh and could provide durable solutions to climate displacement. All of these actions are based on human rights standards. All of these actions can, and should, be supported by the regional and international communities, as well as local and international civil society and NGOs working in Bangladesh. The proposed five key actions for ensuring rights of climate displaced peoples are as follows:

Action One-Climate displacement monitoring mechanism should be implemented across Bangladesh: The Bangladesh Climate Change Strategy and Action Plan (2009) recommends that the Government of Bangladesh: "develops a monitoring mechanism of migration of climate-change-affected people and monitoring of internal as well as external migration". However, despite this recommendation, there still exists no mechanism for monitoring or recording climate displacement across Bangladesh. The Government of Bangladesh should develop and implement a nationwide climate displacement monitoring mechanism to monitor and record all displacement as a result of the effects of climate change. This mechanism could include the ability to register all climate displaced persons, as well as document any support or assistance they are receiving from the Government or other stakeholders.

This comprehensive information can then be utilized nationally, to plan for and implement effective and durable rights-based solutions for all climate displaced persons in Bangladesh. Accurate information on the true scale of climate displacement in Bangladesh is an essential step towards providing effective, rights-based, responses to climate displaced persons. This information could also be shared with other countries, similarly affected by climate displacement, in order to develop global best practice on monitoring and recording climate displacement.

The design of this climate displacement monitoring mechanism can draw on international practice, including the recent joint Government of Samoa and UNDP project entitled 'Human Rights Monitoring of Persons internally displaced by the 2009 Tsunami in Samoa'. The purpose of that study was to monitor and advise on responses to human rights challenges, a well as to bring attention to Internally Displaced Persons (IDPs) as a category of disaster affected persons with a right to protection, appropriate assistance, and active involvement in finding solutions to displacement through return, local integration or relocation.

Action Two-The rights of climate displaced persons should be incorporated into existing climate change law and policy: The Government of Bangladesh has developed a large number of laws and policies relating to climate change vulnerability and adaptation, including: The National Environment Policy (1992);

The Coastal Zone Policy (2005); The National Adaptation Program of Action (2005); The Bangladesh Climate Change Strategy and Action Plan (2009); The National Plan for Disaster Management (2010-2015); The National Land Use Policy (2001); and The Disaster Management Act (2012). However, none of these laws or policies clearly addresses the challenge of climate displacement. It is clear from recent experience that there are considerable gaps and weaknesses in the existing institutional arrangements and existing policies for ensuring the rights of climate displaced persons – including their housing, land and property rights.

An essential step in creating effective responses to climate displacement will be the design and implementation of rights-based laws and policies. Many laws and policies on climate change already exist and it is essential that the rights of climate displaced persons are incorporated into these existing laws and policies. The UN Guiding Principles on Internal Displacement explain the responsibilities of the Government both prior to displacement, during displacement and after displacement. It is essential that all of these responsibilities are incorporated into existing laws and policies. It is equally essential that these laws and policies are drafted in a manner which emphasize respect for the human rights of all climate displaced persons.

It will only be through a concerted effort by the Government as well as from civil society, with the support of the regional and international communities, that effective and durable solutions can be found for the many current and future climate displaced persons in Bangladesh.

Action Three-Distribution of Government Khas land should be effective, transparent and just and take into account the needs of climate displaced persons: Since independence the Government of Bangladesh has enacted a number of laws and policies regarding the distribution of Government *Khas* land. Currently, Article 53 of the Land Management Manual (1991) provides that any landless family is eligible for *Khas* land distribution. *Khas* land is officially State-owned land and is often located in marginal areas along the coast and rivers. The majority of these laws and policies have targeted "landless" persons and families for the grant of *Khas* land. However, these program have met with mixed success due to vested interests illegally occupying *Khas* land, a lack of political will, the inefficiencies in the way local and national administration is organized and the absence of an updated, systematic and universally accepted source of information on land resource availability and land rights.

There is, however, a large amount of agricultural and non-agricultural *Khas* land under the control of the Government and it is clear that this land could play an important part in creating durable solutions for climate displaced persons. It is estimated that State has 3.3 million acres of *Khas* land – of which 25% is agricultural, 50% is non-agricultural and 25% is covered by water bodies. It is essential that the Government take steps to implement an effective, transparent and just program for the distribution of *Khas* land to landless persons - including climate displaced persons. These programs should be rights-based, they should involve the participation of affected communities in their design and the ability to review adverse decisions should be clear and accessible. It is important that decisions about the distribution of Khas land are made on the basis of genuine need, rather than political or personal considerations. Civil Society representatives should also be part of the decision-making panels for *Khas* land distribution. Furthermore, training should be provided to decision makers on climate displacement in Bangladesh and the need to ensure rights-based durable solutions for climate displaced persons.

Article 54 and 56 of the Land Management Manual (1991) provides that persons who are landless as a result of river erosion should be given priority for the allocation of *Khas* land. It is important that this Manual is updated to reflect the current reality that the vast majority of people are and will become landless as a result of a range of adverse effects of climate change, rather than simply prioritizing those people affected by river erosion. The law should give first priority all people affected by the adverse consequences of climate change and natural disasters equally – including victims of tropical cyclones, storm surges, flooding, droughts and landslides.

Action Four-Non-agricultural Khas land should be allocated for climate displaced persons: At present it is only possible for the Government to grant legal title to agricultural land for landless persons (including climate displaced persons). The Government is restricted – under Articles 102 and 103 of the Land Management Manual 1991 – to granting simple leases over non-agricultural land to landless persons. It is clear that domestic land solutions will play an important role in promoting durable solutions for the current and future millions of climate displaced persons in Bangladesh. However, it is equally clear that there is a severe shortage of land in Bangladesh, coupled with dramatic overcrowding in the major cities and slums. For this reason, it is essential that the Government of Bangladesh is able to utilize all land – both agricultural and non-agricultural – in providing solutions to climate displaced persons.

Further, the effects of climate change – including flooding, storm surges, droughts and river erosion - are decreasing the amount of available agricultural land, further emphasizing the need for a policy and legislative change to allow the grant of non-agricultural land to climate displaced persons. It is currently estimated that the State has 1.75 million acres of non-agricultural *khas* Land (50 percent of the total 3.3 million of acres of *Khas* Land). With a change of existing law and policy, the vast majority of this land could be made available to climate displaced persons – with enhanced tenure security – representing an important step towards truly durable solutions to this crisis.

Action Five-Effective return, relocation and rehabilitation programs should be implemented for all climate displaced persons: In line with the Government of Bangladesh's responsibilities under domestic and international law, effective return, relocation and rehabilitation programs should be promptly implemented for all climate displaced persons in Bangladesh. The UN Guiding Principles on Internal Displacement state, in accordance with international human rights and humanitarian law, that the competent authorities (in this case the Government of Bangladesh) have the primary duty and responsibility to facilitate the conditions as well as provide the means for internally displaced persons (including climate displaced persons) to return to their homes or places of habitual residence, or to facilitate and provide the means for their relocation to other parts of the country.

The Guiding Principles also emphasize that special efforts should be made to ensure the full participation of internally displaced persons in the planning and management of their return, or relocation. Currently, there are no comprehensive program in Bangladesh to ensure the effective return of climate displaced persons to their homes or places of habitual residence, nor to facilitate their relocation to other parts of Bangladesh. There are also no comprehensive programs to ensure the effective rehabilitation of climate displaced persons upon return or relocation. Experience has shown that there are many critical livelihood and other challenges to the effective rehabilitation of many climate displaced persons in Bangladesh.

It is essential that such return, relocation and rehabilitation program are designed in a rights-based manner and implemented immediately. The Guiding Principles also note the importance of cooperation by international humanitarian organizations and other actors in assisting with the return or relocation and rehabilitation of climate displaced persons.

Concluding Remarks

Climate displacement in Bangladesh will only worsen as climate change increases the frequency and intensity of the natural hazards that are already leading to displacement across Bangladesh. One of the most catastrophic effects of climate change is climate displacement – the displacement of persons from their homes and lands as a result of the effects of climate change. It is essential that effective and durable solutions to this growing crisis are developed and implemented immediately. It is equally essential that different stakeholders in Bangladesh and in the regional and international communities contribute to ensure the implementation of a truly effective institutional framework – through financial and technical cooperation, as well as ongoing monitoring and evaluation of the legal and policy responses to climate displacement. Hence The Bangladesh Housing Land and Property (HLP) Rights Initiative has been working as catalyst among the various stakeholders particularly targeting Government of Bangladesh as final authority to take durable initiative for climate displacement in Bangladesh.

At present there is not yet a truly coordinated and effective institutional framework in place to respond to and plan for this challenge. Conclusion can be drawn with the following seven steps recommendations[11] that could be taken by the Government of Bangladesh and by national, regional and international stakeholders, in order to create and implement a truly effective and coordinated response to climate displacement in Bangladesh:

Recommendations:

1. **Undertake a comprehensive review of existing national laws and policies:** The Government of Bangladesh, alongside key stakeholders, should systematically review and assess all of the existing national laws and policies that touch on the question of climate displacement in Bangladesh. The existing legal and policy framework in its current form is not sufficient to provide comprehensive protection to climate displaced persons in Bangladesh. A systematic review and analysis of the existing normative framework will identify key protection gaps and

will be an important first step towards the design and implementation of truly effective rights-based laws and policies.

2. Design and implement rights-based national laws and policies: After the review of existing laws and policies is completed, the Government – in consultation with key stakeholders, including climate affected communities - should take immediate steps to design and implement rights-based laws and policies that ensure the effective protection of climate displaced communities in Bangladesh. This may involve the enactment of an entirely new, stand-alone law on climate displacement in Bangladesh. This process should draw on international best practice, as well as extensive and genuine consultation with national stakeholders, including climate-affected communities. The end result should be a comprehensive, rights-based framework for responding to and preparing for climate displacement in Bangladesh. The framework should focus on preventing displacement, responding to displacement where it occurs and providing rights-based durable solutions (including return, local integration and relocation) where possible.

3. Undertake a review and re-design of the existing institutional framework: The Government should review and assess the existing institutional framework relevant to climate displacement. At present there are a large number of Government Ministries, Agencies and Departments with direct or indirect responsibility for climate displacement. This plethora of institutions, without clear lines of responsibility, reduces accountability and clarity in the Government's response to climate displacement. Drawing on the results of this writings, the Government should implement an effective and accountable institutional framework – with clear lines of responsibility – for responding to climate displacement. This should include mechanisms to increase coordination and communication between national policy and decision makers and local level officials who are at the front-line of implementing national policy and laws on climate displacement. This institutional framework should include sufficient training on climate displacement for key government officials and representatives, as well as mechanisms to improve transparency in the implementation of policies and laws, as well as donor and budgetary funds.

4. Design and implement domestic land solutions: Climate displacement laws and policies should provide for domestic land solutions for climate displaced persons in Bangladesh. At present, international law does not provide a clear right of asylum or other forms of protection to climate displaced persons who cross international borders. In the absence of international legal protection, the first priority for rights-based solutions should lie within the national borders of Bangladesh. Despite having an already high population density, studies have revealed that there are large tracts of Government *Khas* land that are available and suitable for the relocation of climate-displaced communities.[12] The Government should ensure that suitable *Khas* land is made available for the relocation of climate displaced communities, including through reviewing and improving the existing programs for the distribution of *Khas* land to landless persons in Bangladesh.

5. **Increase coordination and knowledge sharing among international and regional donors:** There are a large number of regional and international donors funding a variety of different programs relevant to climate displacement in Bangladesh. In order to ensure efficiency and to avoid duplication of efforts and resources, the international and regional donor communities should create a Donor Working Group on Climate Displacement in Bangladesh. This group could meet regularly to discuss and coordinate the use of donor funds and to share information on successes and challenges associated with resolving climate displacement in Bangladesh. This working group could also benefit from regular roundtable discussions and presentations on climate displacement from international and Bangladesh experts, utilizing the important academic and research expertise that is being developed in Bangladesh. This working group could facilitate more effective communication on climate displacement to capitals and headquarters in order to ensure sufficient financial and technical cooperation on this issue.

6. **Create a UN and NGO working group on climate displacement:** There are a large number of international NGOs and UN agencies working directly and indirectly on different aspects of climate displacement. It is recommended that a UN and NGO working group on climate displacement in Bangladesh be established. Similarly to the Donor Working Group, this group could meet regularly to share knowledge on the situation of climate displacement as well as discuss successes and challenges in seeking to resolve climate displacement in a rights-based manner in Bangladesh. This group could also benefit from roundtable discussions and presentations by Bangladesh and international experts on climate displacement. This group could provide an important mechanism for providing UN and other

headquarters with essential information and developments on climate displacement in Bangladesh.

7. **Improve coordination and communication between the Government and civil society:** The Government should take steps to improve coordination between Government institutions and civil society in Bangladesh, including grass roots and national organizations, academic and other research institutions. These organizations have a wealth of up to date information and understanding of the situation of climate displacement, as well as important links with remote climate affected communities. This could include the creation of Government and civil society working and coordination groups at the national and District levels. These working groups could meet to share information on the latest developments on climate displacement, as well as to ensure that national laws and policies are being effectively implemented at the local level.

However, it is matter of hope that Dr. Hasan Mahmud, Member of Parliament and former Minister of Environment and Forest of Bangladesh Government said that "a comprehensive Climate Change Policy is now demand of time, which can cover all climate changes issues including the rights of climate displaced persons of Bangladesh. I, myself, will take necessary initiative for this policy as a present chairman of Parliamentary standing Committee for Ministry of Environment and Forest" spoke as chief guest in a Round Table Discussion on Climate Displacement in Bangladesh issue recently organized by YPSA at Dhaka.

References

[1] Displacement Solutions, 2012. Climate Displacement in Bangladesh: The Urgent Need for Housing, Land and Property Solutions.

[2] IPCC Fifth Assessment Report, 2014. Working Group I.

[3] Germanwatch, 2013. Global Climate Risk Index 2014: Who Suffers Most from Extreme Weather Events? Weather-Related Loss Events in 2012 and 1993 to 2012

[4] Ahmed, M. and S. Suphachalasai., 2014. Assessing costs of climate change and adaptation in South Asia, Mandaluyong City, Philippines: Asian Development Bank.

[5] Gaim Kibreab, 2010. 'Climate Change and Human Migration: A Tenuous Relationship?' *Fordham Environmental Law Review*

[6] Comprehensive Disaster Management Program (CDMP II), 2014. "Trend and Impact Analysis of Internal Displacement due to Impact of Disaster and Climate Change", Ministry of Disaster Management and Relief, Dhaka.

[7] Tanner, T.M.; Hassan, A.; Islam, K.M.N.; Conway, D.; Mechler, R.; Ahmed, A.U. and Alam, M., 2007. ORCHID: Piloting Climate Risk Screening in DFID Bangladesh, Summary Research Report, Institute of Development Studies, University of Sussex, UK.

[8] Akter, T., 2009. Climate Change and flow of environmental displacement in Bangladesh, Unnayan Anneshan-The innovators, Dhaka, Bangladesh.

[9] Displacement Solutions, 2012. Climate Displacement in Bangladesh: The Urgent Need for Housing, Land and Property Solutions.

[10] DS and YPSA, 2014. Climate Displacement in Bangladesh: Stakeholders, Laws and Policies - Mapping the Existing Institutional Framework.

[11] Ibid.

[12] Scott Leckie (ed), 2014. Land Solutions to Climate Displacement, Routledge/Earthscan.